Weather and Climate

Written by
Judy Langthorne
and
Gaye Conroy

Illustrated by
Robert Wheeler

Wayland

Books in the series

Clothes and Costumes Landscapes
Conservation Resources
Houses and Homes Water
Journeys Weather and Climate

This edition published in 1994 by
Wayland (Publishers) Ltd

First published in 1992 by
Wayland (Publishers) Ltd
61 Western Road, Hove
East Sussex, BN3 1JD, England

Series editor: Mandy Suhr
Designer: Jean Wheeler

British Library Cataloguing in Publication Data

Conroy, Gaye
Weather and climate.–(Starting geography)
I. Title   II. Langthorne, Judy   III. Series
551.5

HARDBACK ISBN 0-7502-0315-3

PAPERBACK ISBN 0-7502-0614-4

Typeset by DP Press, Sevenoaks, Kent
Printed in Italy by Rotolito Lombarda, S.p.A., Milan

Contents

The words that appear in **bold** are explained in the glossary.

Think about the weather

Water, wind and sunshine together make our weather.

Wherever we go and whatever we do, the weather affects us all. The type of home we live in, the clothes we wear, even the food we eat depends on the weather.

Would you eat an ice-cream on a freezing cold day?

The weather affects the amount of water there is for growing crops and keeping ourselves, as well as plants and animals, alive.

Different countries have very different kinds of weather.

What is the weather like today?

Shapes in the sky

Look up at the sky. The shapes of the clouds and the colour of the sky will give you a clue about the weather to come.

Clouds are made from millions of tiny drops of water.

The heat of the sun causes these drops to rise up from the seas and from rivers.

You cannot see the drops until they cool down and collect together into a cloud.

Sometimes the weather is foggy.
Fog is really a very low cloud,
down near the ground.
In Britain, there are lots of foggy
days in the winter. The fog makes
it difficult to see, especially on
the roads.

Look at the sky. Can you guess what the weather will be like today?

Where's my umbrella?

Some people love the rain, others hate it. But we could not survive without water.

The water we use for washing, cooking and drinking all started off as rain-water.

In some countries it rains quite often. Other countries have very little rain.

Some countries, like India, have a rainy season when it rains very heavily for several months.

sun

clouds

rain

water vapour rises

lake

sea

Rain comes from the clouds and falls on the earth making our rivers and streams, ponds and lakes. This water gets heated by the sun. Tiny drops of water rise up and cool to make clouds. As these drops cool down even more, they get heavier and fall as raindrops. This is called the rain cycle.

Activity

Make a poster to show all the different ways that you use water.

Fun in the sun

Some countries, like Australia and parts of the USA, have lots of hot, sunny days.

Some other countries may have just a few.

The children in this picture have gone to a warm, sunny country for their holidays.

Without the sun the world would be a very gloomy place.

The sun gives us light and heat. This helps all living things to grow. Some crops, like these oranges from Florida in the USA, and bananas from Jamaica, will grow only in hot, sunny places. ▶

Farmers from hot countries sell their crops to countries that are not so sunny. The crops are **transported** by ship or aeroplane across the world.

What do you like doing on hot sunny days?

Activity

Investigate where different fruit and vegetables come from. Can you find these countries on a map?

Pineapple

Pepper   Aubergine   Banana

Grapes

Orange

Avocado

Windy days

What can you hear and feel but not see? The wind!

The wind is really moving air. It can come from many different directions.

Whether the wind is cold or warm depends on where it has come from. Warm winds have travelled across hot **deserts** in the south. Cold winds have travelled across cold, icy places in the north.

A **weather vane** can tell us which direction the wind is coming from.

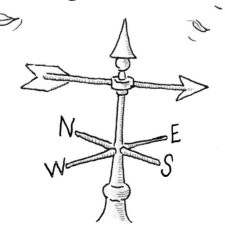

Sometimes the wind is very strong. It can blow over dustbins or even knock down trees. These strong winds can bring rain and violent storms.

The wind can be used to turn the sails of a **windmill** to make electricity, pump water or grind corn.

Look for weather vanes, **windsocks** and windmills around where you live.

The power of the wind turns the blades of these wind machines to make electricity.

Footprints in the snow

When it is very cold, the water droplets in the clouds freeze and fall as snowflakes.

Some places, like Greenland, are covered with ice and snow all year round. Other countries, like Kuwait, never have snow.

Do you like snowy weather? Skiing, tobogganing and snowball fights are all fun, but snow can also be dangerous.

Snow on the roads can make travelling very difficult.

Snow is blown by strong winds into big piles, called snow-drifts. Snowstorms are often known as blizzards. In a blizzard, roads can become blocked and airports close. **Snowploughs** are brought out to clear the snow-drifts away. ▶

In countries that have lots of snow, people may use different ways of travelling. These children are wearing warm and waterproof clothes. They are using skis to go from place to place. ▼

All around the world

Have you ever wondered why some countries are hot and some are cold? Why some are very dry and others have lots of rain? This is caused by their climate. This is the pattern of weather that a place has throughout the year.

The climate depends on how near a country is to the Equator. The Equator is an imaginary line, halfway between the **North Pole** and the **South Pole**. Around the Equator are the hottest parts of the world. ▼

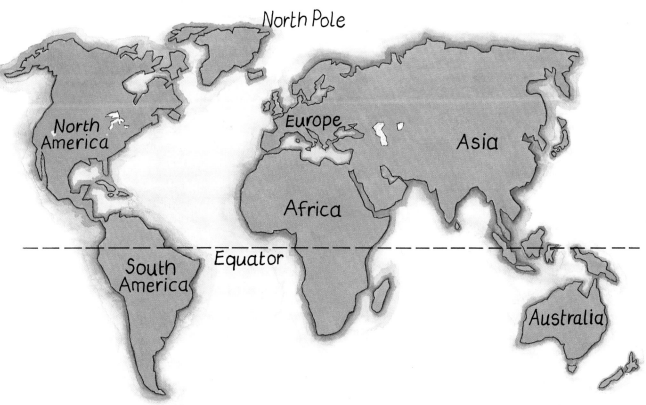

The coldest parts of the world are the North and South Poles. Countries that are near to these have very cold, snowy climates.

Some countries, like the USA, are so big that different parts of the country may have different climates.

Activity

Talk to people you know who have lived in climates different from your own. Collect postcards from countries that have different climates.

Hot places

Countries that are near to the Equator have hot weather all year round. This is called a **tropical climate**.

Some of these countries have large rain forests which make the climate very **humid**. This means that the air is very wet. ▼

Many plants and animals that live in the rain forests cannot be found anywhere else.

These children live in the rain forests of Brazil. ▼

Countries around the Equator that do not have rain forests are also very hot, but the air is drier.

Some countries have large areas of desert land which is very dry. Plants and animals that live there must be able to survive with very little water. ▼

In these hot countries, many buildings, and even some cars, have **air-conditioning** which cools the air.

These people in Morocco, North Africa, are wearing clothing that covers their skin and protects them from the hot sun and dust. ▼

Cold places

North Pole

South Pole

Find a **globe** of the world. If you look at the top you will find the **Arctic**. This is the area of land and frozen sea around the North Pole.

At the bottom of the globe you will find the **Antarctic**. This is the area around the South Pole.

The two poles are the parts of the world furthest from the Equator. Most of the land is covered with snow nearly all year round and the sea freezes. It is cold every day.

Animals and plants that live in these cold places have to be able to survive in very cold weather. Some animals, like polar bears, have thick furry coats to keep them warm. ▶

Countries that are near to the poles have a cold, icy climate called a **polar climate**.

These children live in northern Canada. It is so cold, they can play ice hockey in the street. ▼

Changing weather

Some countries are not close to the Equator or close to the poles. This means that the weather is never really hot or really cold but changes all year round. It can be warm and sunny or cold and frosty depending on the time of year. This is known as a **temperate climate**.

Homes in these countries have to stand up to all these different types of weather.

They often have sloping roofs so that the rain will drain off. Most modern houses have heating which can be turned on or off depending on the weather. ▼

sloping roof

electric fire

radiators

double glazing

The clothes people wear also change at different times of the year. In the warmer months of spring and summer their clothes will be cooler. ▶

▲
In winter and autumn, when the weather is cold and wet, their clothes will be warm and waterproof.

What kind of climate would you like to live in?

Seasons come and seasons go

These two pictures show the same town in winter and in summer.

Why do you think the weather is not the same all year round? The weather changes from one season to the next.

The seasons change because the Earth is in orbit around the Sun. This means that is spins around the Sun.

The Earth takes a year to travel all the way around. The Earth is tilted so as it orbits the Sun, some parts are near to the Sun and others are further away, and so the seasons change.

Countries with a temperate climate have four seasons. Winter is cold, wet and windy and spring has warmer, showery weather. Summer has hot, sunny days and in autumn the weather begins to cool down again.

Not all countries have the same seasons. Countries with a polar climate have a cool summer for half of the year, and a very cold winter for the other half.

Many countries with a tropical climate stay hot all year round but have a dry season and a wet season. ▶

Australia's seasons happen at opposite times of the year to those in Britain. These children are celebrating Christmas by the beach. ▼

Weather disasters

Many children around the world have been frightened by storms. These are given different names in different countries. Sometimes they are called typhoons or tornadoes, sometimes hurricanes or cyclones.

In 1987, a hurricane blew across southern England. It blew down trees and blew roofs off houses. In Jamaica, they also have hurricanes. Some are given names like 'Hurricane Hugo'. This hurricane destroyed many homes.

Sometimes there is not enough rain and the land becomes very dry. This is called a **drought**. ▶

Crops are not able to grow, so there may not be enough food to feed everyone. Because of the shortage of water there may not be enough to drink.

◀ When it rains for many days, the land can become **flooded**. In some countries, people build their houses in special ways to stop them being flooded. But sometimes even this is not enough.

Find out about other kinds of dangerous weather.

What will the weather be like tomorrow?

All over the world people need to know what the weather is going to be like. Sailors need to know if there are going to be **gales** at sea. Farmers need to know if they should harvest their crops.

Sometimes special warnings are given by the **weather forecasters** about dangerous weather conditions, like fog, hurricanes, floods or snow blizzards.

Warnings can help people to protect themselves against dangerous weather.

▲
This man is giving a weather forecast on television.

Special instruments on aeroplanes, balloons, ships and **satellites** send information about the weather back to local weather stations in countries all over the world.

This scientist is collecting information about weather in the Arctic. ▶

Activity

Look through newspapers and cut out weather charts. Compare weather around the world.

Listen to weather reports on the radio and television.

Glossary

Air-conditioning A way of keeping the air inside a car or building cool.

Antarctic The area around the South Pole.

Arctic The area around the North Pole.

Deserts Large areas of very dry, often sandy, land.

Drought A long period of dry weather, creating a water shortage.

Flood A large amount of water covering land that is normally dry. It is often caused by very heavy rain.

Gales Very strong winds.

Globe A sphere showing the map of the world.

Humid Warm and damp weather.

North Pole The part of the earth's surface that is the most northerly.

Polar climate A pattern of cold, icy weather in countries near to the poles.

Satellites Machines sent into space to collect information.

Snowplough A vehicle used to clear snow from the roads.

South Pole The part of the earth's surface that is the most southerly.

Temperate climate A pattern of weather that changes all year round depending on the season.

Transported When goods or people are transported they are moved from one place to another.

Tropical climate A pattern of hot weather that can be dry or humid. This occurs in countries that are near to the Equator.

Weather forecasters People who collect information about the weather to come.

Weather vane An instrument that is blown by the wind and points to show which direction the wind is coming from.

Windmill A building or object that has sails which turn as the wind blows. This wind power is used to grind corn, pump water or make electricity.

Windsocks Sock-shaped bags which fill with air when the wind blows and show the direction of the wind.

Finding out more

Books to read

Will it Rain Today by Althea (Dinosaur Publications, 1988)

Life in Deserts: A Jump Ecology Book by Lucy Baker (Franklin Watts, 1990)

Life in the Polar Lands: A Jump Ecology Book by Monica Byles (Franklin Watts, 1990)

Let's Look at Rain by Jacqueline Dineen (Wayland, 1988)

BBC Factfinders – Weather by Tony Potter (BBC Enterprises Ltd, 1989)

You may find the following series useful:

What's the Weather Like? (Wayland, 1988)

The Weather (Wayland, 1992)

Picture acknowledgements

B & C Alexander 14, 15 (top), 17 (left), 20, 21 (top), 21 (bottom), 29; Bruce Coleman 4, 5 (top), 25 top); 24 (right), 24 (left); Christine Osborne 17 (right), 19 (left), 19 (right), 25 (bottom); Ecoscene 27 (top); Frank Lane Picture Agency 7; Lupe Cunha 10, 18 (bottom); Photri 8, 11, 13; Tony Stone *front cover*, 23 (top), 23 (bottom), 26; Wayland Picture Library 5 (bottom), 16, 18, 28; Zefa 6, 12.

Index